CW01309900

Materialism Theory and the Concept of Unit Particles

A New Framework for Fundamental Physics, Gravity, and the Structure of the Universe

Manoranjan ghoshal

Materialism Theory and the Concept of Unit Particles
A New Framework for Fundamental Physics, Gravity, and the Structure of the Universe
© 2025 **Manoranjan Ghoshal**
All rights reserved.

No part of this book may be reproduced, stored in a retrieval system, or transmitted in any form or by any means—electronic, mechanical, photocopying, recording, or otherwise—without the prior written permission of the author, except in the case of brief quotations embodied in critical articles and reviews.

Published by:
Manoranjan ghoshal
First Edition: 2025

For inquiries, permissions, or bulk purchases, please contact:
ghoshal.manoranjan@gmail.com

Book Description

What if all matter, from electrons to galaxies, is built from a single, indivisible particle?

In *Materialism Theory and the Concept of Unit Particles*, physicist **Manoranjan Ghoshal** introduces **Unit Particle Theory (UPT)**, a revolutionary framework proposing that **the Unit Particle—the tiniest, fundamental, and indestructible entity—forms all known matter**. With a **fixed mass and volume**, Unit Particles combine in **exact integer ratios** to create electrons, protons, and neutrons, laying the foundation of everything in existence.

This book explores:
1. **The Unit Particle as the true fundamental building block of reality**
2. **How electrons, protons, and all particles arise from integer-grouped Unit Particles**
3. **Gravity as an effect of Unit Particle density, not spacetime curvature**
4. **New insights into dark matter, dark energy, and the structure of the universe**
5. **How UPT challenges the Standard Model and General Relativity**

Blending **scientific depth with bold theoretical insights**, *Materialism Theory and the Concept of Unit Particles* offers a **new path toward understanding the fundamental nature of the universe**. If you seek the **true essence of matter**, this book provides a groundbreaking perspective on modern physics.

Acknowledgement

I extend my heartfelt gratitude to all those who have inspired and supported me in developing **Unit Particle Theory (UPT)**.

I sincerely thank my mentors, colleagues, and fellow researchers for their valuable insights and discussions that shaped this work. My deep appreciation goes to my family and friends for their unwavering encouragement and patience throughout this journey.

Lastly, I acknowledge the great minds of physics—past and present—whose relentless pursuit of truth has paved the way for new ideas. This book is a small step toward that endless quest for understanding the universe.

Manoranjan Ghoshal

Table of Contents

Preface

Introduction

Chapter 1: Foundations of Materialism in Physics
Explores classical and modern materialist thought, tracing the philosophical lineage that leads to Unit Particle Theory.

Chapter 2: The Concept of the Unit Particle
Introduces the Unit Particle as the indivisible, eternal fundamental building block of all matter and energy.

Chapter 3: Structure of Matter According to UPT
Describes how combinations of Unit Particles form fundamental particles like electrons, protons, and neutrons.

Chapter 4: Forces and Fields in UPT
Explains how all known forces arise from gradients, interactions, and waves in the Unit Particle field.

Chapter 5: Space, Time, and Motion
Redefines time as the evolution of configurations within the Unit Particle substrate and offers a new interpretation of space and motion.

Chapter 6: Light and Electromagnetic Phenomena
Explores wave propagation, Redshift, and electromagnetic interaction through the lens of Unit Particle behaviour.

Chapter 7: Cosmological Implications of UPT
Applies the theory to the origin, structure, and fate of the universe—offering alternatives to the Big Bang, dark matter, and dark energy.

Chapter 8: Experimental Verification and Future Research Directions
Details testable predictions, laboratory setups, and paths for further exploration of UPT.

Chapter 9: Philosophical Reflections and the Future of Scientific Thought
Discusses UPT's implications for determinism, causality, metaphysics, and the philosophy of science.

Supplementary Material

Appendix A: Mathematical Formulations of UPT
Presents equations governing Unit Particle dynamics, field interactions, and emergent properties.

Appendix B: Glossary of Terms
Defines key concepts, technical language, and unique terminology used in the book.

Bibliography
Cites foundational works and relevant academic sources that support or contrast with UPT.

Final Compilation: A Unified Vision of Matter and Physics

Preface

Throughout history, physicists and philosophers have sought the fundamental essence of reality—the indivisible building block from which everything emerges. From **Democritus' atomic theory** to the **Standard Model of particle physics**, our understanding has evolved, yet many unanswered questions remain. **What is the true foundation of matter? Why do fundamental particles have specific properties? Can we unify all forces under a single framework?**

In this book, I introduce **Unit Particle Theory (UPT)**—a new approach that challenges conventional physics by proposing a **single, indivisible particle as the foundation of all matter and forces.** This **Unit Particle** is the **smallest, undivided, and indestructible entity with a fixed mass and volume**. It cannot be created or destroyed, yet it is responsible for forming all known particles, including **electrons, protons, and neutrons, through precise integer combinations**.

This perspective eliminates the need for a complex set of elementary particles and quantum fields, instead suggesting that the entire **universe is built upon a fundamental, integer-based structure**. By understanding how **Unit Particles interact, cluster, and organize**, we gain new insights into:

1. **The structure of matter** and the composition of elementary particles
2. **Gravity as an emergent effect** rather than a fundamental force

3. **A deterministic approach to quantum mechanics**, resolving key paradoxes
4. **The nature of space and time**, reinterpreted through the lens of Unit Particle interactions
5. **The mysteries of dark matter and dark energy**, explained through Unit Particle density variations

Modern physics, despite its successes, still faces deep contradictions—most notably, the **incompatibility between Quantum Mechanics and General Relativity**. If we are to truly understand the universe at its most fundamental level, we must be open to **reconsidering the basic assumptions of physics**. Unit Particle Theory offers an **alternative foundation**, one that provides a **logical, consistent, and unified picture of reality**.

I invite scientists, researchers, students, and thinkers to explore this **new vision of the physical universe**. If we are to uncover the deepest truths of existence, we must be willing to challenge the established frameworks and **seek answers at the most fundamental level of matter itself—the Unit Particle**.

Manoranjan Ghoshal

Introduction

1.1 The Search for a Fundamental Particle

For centuries, scientists and philosophers have sought the most basic building block of the universe—the **fundamental particle** from which all matter is formed. Ancient Greek philosophers such as **Democritus** imagined tiny, indivisible "atoms" as the foundation of reality, while modern physics has introduced an increasingly complex array of elementary particles, quantum fields, and mathematical constructs.

Despite the successes of **the Standard Model of particle physics**, which describes the fundamental forces and particles of nature, it remains incomplete. It cannot explain **the true origin of mass**, **the nature of gravity**, or **the mysteries of dark matter and dark energy**. Furthermore, the fundamental particles in the Standard Model—quarks, leptons, bosons—do not seem to be the ultimate indivisible entities. Instead, they appear as different states of underlying energy fields, lacking a truly foundational nature.

This book presents an alternative framework: **Unit Particle Theory (UPT)**. It proposes that there exists a **single, fundamental, indivisible particle—called the Unit Particle—from which all matter, forces, and even space emerge.**

1.2 Defining the Unit Particle

The **Unit Particle** is the smallest, most elementary entity in nature. It possesses the following key properties:

1. **Indivisibility** – The Unit Particle cannot be broken down into smaller components. It is the true fundamental particle of nature.
2. **Fixed Mass and Volume** – Every Unit Particle has an identical, unchanging mass and occupies a definite volume in space.
3. **Integer-Based Structure** – All known particles (electrons, protons, neutrons) are composed of **exact integer numbers of Unit Particles**.
4. **Conservation** – Unit Particles cannot be created or destroyed, ensuring the fundamental conservation of matter at the deepest level.

According to **Unit Particle Theory**, all observed elementary particles in the Standard Model are not truly fundamental but rather **conjugations of multiple Unit Particles in precise integer ratios**. For example:

1. An **electron** consists of **X Unit Particles**.
2. A **proton** consists of **Y Unit Particles**, where **Y > X**.
3. A **neutron** consists of **Z Unit Particles**, where $Z \neq X, Y$ but still an integer multiple of Unit Particles.

This theory provides a **simpler and more unified explanation** of matter, replacing the need for multiple distinct "fundamental" particles with just **one**: the **Unit Particle**.

1.3 Why the Standard Model is Incomplete

The **Standard Model of particle physics** has been incredibly successful in explaining the fundamental forces (except gravity) and predicting the behavior of elementary particles. However, it has major shortcomings:

1. **Too Many Fundamental Particles** – Why does nature require multiple types of quarks, leptons, and bosons instead of one true building block?
2. **No Explanation for Gravity** – The Standard Model does not include gravity, making it incompatible with **General Relativity**.
3. **Dark Matter and Dark Energy Are Unexplained** – Observations suggest that 95% of the universe consists of unseen matter and energy that the Standard Model does not account for.
4. **Arbitrary Constants** – Particle masses, force strengths, and other parameters are inserted **manually**, rather than emerging from a deeper principle.

Unit Particle Theory addresses these problems by proposing that all particles arise from combinations of a single fundamental entity—the

Unit Particle—eliminating the need for multiple arbitrary constants and unexplained forces.

1.4 A New Approach to Gravity and Space-Time

One of the most revolutionary aspects of **Unit Particle Theory** is its **reinterpretation of gravity**. In contrast to Einstein's **General Relativity**, which describes gravity as a curvature of spacetime, **UPT suggests that gravity is an effect of Unit Particle density variations.**

1. In high-density regions (like stars and planets), **Unit Particles cluster together**, creating an attractive force we perceive as gravity.
2. In low-density regions (deep space), **this attraction weakens**, leading to cosmic expansion.

This idea also provides a **new explanation for dark matter and dark energy**:

1. **Dark Matter** may be composed of an **unobservable state of Unit Particles**, influencing galaxies without interacting via electromagnetic forces.
2. **Dark Energy** may result from **density fluctuations of Unit Particles**, causing the universe's accelerated expansion.

By treating gravity as an **emergent effect of Unit Particle arrangements**, UPT offers a potential **unification of quantum mechanics and gravitational physics**—something modern physics has struggled to achieve.

1.5 The Purpose of This Book

This book presents **Unit Particle Theory** as a revolutionary new perspective in physics. It seeks to:

1. **Introduce the concept of the Unit Particle** and how it replaces the need for multiple elementary particles.
2. **Show how all fundamental particles (electrons, protons, etc.) are integer-based groupings of Unit Particles.**
3. **Explain gravity as an emergent effect of Unit Particle density.**
4. **Provide a new framework for understanding dark matter, dark energy, and space-time itself.**
5. **Challenge the Standard Model and General Relativity by offering a simpler, unified theory of matter and forces.**

This book is intended for **physicists, students, and anyone interested in fundamental physics**. While some sections include technical details, the core concepts are presented in a way that is accessible to all **who seek to understand the deepest truths of reality**.

If **Unit Particle Theory is correct**, it will revolutionize our understanding of the universe—perhaps even leading to new breakthroughs in **energy, space travel, and quantum computing**.

Let us now embark on this journey to explore the **smallest yet most powerful entity in existence: the Unit Particle**.

Chapter 1: The Unit Particle – The True Fundamental Entity

1.1 The Need for a Single Fundamental Particle

Modern physics describes nature in terms of multiple fundamental particles—quarks, leptons, and bosons—each with different properties and interactions. The **Standard Model** successfully explains these particles and forces, but it does not tell us **why** these particles exist or **what they are fundamentally made of**.

Unit Particle Theory (**UPT**) proposes a **simpler and more fundamental explanation**:

1. **There exists only one truly fundamental particle**—the Unit Particle (UP).
2. **All known particles, including electrons, protons, and neutrons, are composed of exact integer multiples of Unit Particles.**
3. **Unit Particles cannot be created or destroyed**, making them the true **indivisible essence of reality**.

This idea **eliminates the need for multiple types of elementary particles** and suggests that the universe operates on a **universal integer-based structure**.

1.2 Defining the Unit Particle

The **Unit Particle (UP)** is proposed as the **smallest, indivisible, and most fundamental entity in the universe**. It has the following key properties:

1.2.1 Indivisibility and Permanence
1. **The Unit Particle is truly fundamental**—it cannot be divided into smaller parts.
2. Unlike quarks or energy waves, it has a definite **physical existence**.
3. It is **eternal**, meaning it cannot be **created or destroyed** by any process.

1.2.2 Fixed Mass and Volume
1. Every **Unit Particle** has the **same fixed mass** and occupies an **exact volume in space**.
2. This ensures **consistency** in how Unit Particles combine to form larger particles.
3. The **mass of any fundamental particle (electron, proton, neutron, etc.) is simply the sum of the masses of its Unit Particles**.

1.2.3 Integer-Based Composition of Matter
1. Every particle in nature is made of **a specific integer number of Unit Particles**.
2. If an **electron** contains **X Unit Particles**, a **proton** must contain **Y Unit Particles**, where **Y > X**.
3. This explains **why particle masses are discrete and not continuous**.

The concept of integer-based composition suggests a **new kind of mathematical order in nature**, simplifying the complex models used in modern physics.

1.3 How Unit Particles Form Fundamental Particles

The **electron, proton, and neutron** are not truly elementary but instead **structured formations of Unit Particles**.

1. **Electrons** are made of **X Unit Particles**, where X is a fixed, positive integer.
2. **Protons** are composed of **Y Unit Particles**, where Y is another integer, larger than X.
3. **Neutrons** contain **Z Unit Particles**, where Z is distinct from X and Y but still an integer.

This integer-based structure removes the need for concepts like **quarks and gluons**, replacing them with **precise, quantifiable arrangements of Unit Particles**.

1.3.1 Why Are Some Particles Charged and Others Neutral?

Unit Particle Theory suggests that **charge emerges from the spatial arrangement of Unit Particles**:

1. **Electrons**: If an arrangement of Unit Particles follows a certain structure, it **creates an electric charge**.
2. **Protons**: A different structure of Unit Particles generates an **opposite charge**.
3. **Neutrons**: If Unit Particles arrange in a way that balances charges, the resulting particle is **neutral**.

Instead of relying on abstract "charge carriers," UPT explains **electric charge as a fundamental property of Unit Particle organization.**

1.4 The Role of Unit Particles in Atomic Structure

Atoms are **not mysterious quantum objects**—they are structured **collections of Unit Particles.**

1.4.1 Atoms as Structured Groups of Unit Particles

1. Atoms are made of **electrons, protons, and neutrons**, each composed of **Unit Particles.**
2. The **total number of Unit Particles in an atom** determines its **mass and stability.**
3. The **arrangement of electrons (which are made of Unit Particles) around the nucleus** determines chemical properties.

1.4.2 Molecules and Macroscopic Matter

1. When atoms combine to form **molecules,** they are simply **higher-level arrangements of Unit Particles.**
2. This means that **everything in the universe**—solids, liquids, gases, and even energy fields—is ultimately composed of structured Unit Particles.

Thus, **matter is not a collection of independent particles, but a structured organization of the same fundamental Unit Particles.**

1.5 Implications of the Unit Particle Theory

Unit Particle Theory offers a radical **reinterpretation of fundamental physics**:
1. **Replaces the Standard Model's complex set of elementary particles with a single fundamental entity—the Unit Particle.**
2. **Explains why particles have specific masses and charges based on integer-based structures.**
3. **Eliminates the need for quarks, gluons, and multiple fundamental forces.**
4. **Proposes a deterministic, structured foundation for physics, removing quantum randomness.**
5. **Paves the way for new insights into gravity, dark matter, and space-time.**

Here are some additional explanations to further clarify key concepts in **Unit Particle Theory (UPT):**

1.6 The Fundamental Nature of the Unit Particle

One of the most pressing questions in physics is: **What is the ultimate building block of reality?** The Standard Model of particle physics describes many "fundamental" particles, but these particles **appear to have internal structure or arise from more fundamental fields**. This suggests that they are not truly elementary.

Unit Particle Theory (UPT) argues that all observed particles emerge from a single, indivisible entity—the Unit Particle. Unlike quarks, leptons, or energy waves, the Unit Particle has a **definite physical presence**, making it the **true fundamental unit of nature**.

1.6.1 What Makes the Unit Particle Unique?

1. **Indivisible** – Unlike protons (which are made of quarks) or quarks (which may arise from deeper quantum fields), the **Unit Particle cannot be split**.
2. **Permanent** – Unlike photons or virtual particles, **Unit Particles cannot be created or destroyed**.
3. **Structured Organization** – While individual Unit Particles remain **unchanged**, their **arrangement determines the properties of all observed matter**.

This perspective challenges conventional physics, which assumes that the universe operates using **multiple types of fundamental particles**. Instead, **UPT proposes a single, universal entity that builds all known forms of matter**.

1.7 How Unit Particles Form Different Particles

If **all matter is made of Unit Particles**, how do we explain the **variety of observed particles**? The answer lies in **how Unit Particles cluster together**.

1.7.1 Electron Formation

1. The **electron**, often considered a fundamental particle, is actually **a structured grouping of X Unit Particles**.
2. This structure determines its **mass, charge, and wave-like behavior**.

3. The **motion of Unit Particles within an electron** explains **why electrons behave as both particles and waves.**

1.7.2 Proton Formation
1. A **proton** consists of **Y Unit Particles, where Y > X.**
2. The way these **Unit Particles are arranged** determines the **proton's charge (+1) and stability.**
3. Instead of needing **quarks and gluons**, UPT suggests that **protons arise directly from a specific integer combination of Unit Particles.**

1.7.3 Neutron Formation
1. A **neutron** contains **Z Unit Particles, where Z is a unique integer value.**
2. Unlike the proton, the neutron's **Unit Particle arrangement results in a neutral charge.**
3. This explains why **neutrons are stable inside an atomic nucleus but decay when isolated**—their Unit Particle configuration becomes unstable outside the nucleus.

1.7.4 The Ratio of Unit Particles in Matter

UPT suggests that there is a **fixed, integer relationship between the number of Unit Particles in different particles:**

 X Unit Particles = 1 Electron
 Y Unit Particles = 1 Proton (where Y > X)
 Z Unit Particles = 1 Neutron (where Z is different from X and Y but also an integer)

These integer relationships suggest that the universe follows a **structured, mathematical order** rather than relying on random quantum fluctuations.

1.8 The Role of Unit Particles in Charge and Mass

1.8.1 What Determines Charge?

In modern physics, **charge** is an intrinsic property of elementary particles, but its true nature remains unexplained. UPT proposes that:

1. **Charge arises from the geometric arrangement of Unit Particles.**
2. If Unit Particles align in a particular way, they create **a negative charge (electron).**
3. A different structural arrangement leads to **a positive charge (proton).**
4. A **balanced, symmetric arrangement results in no charge (neutron).**

This means that **charge is not a separate entity—it emerges naturally from the configuration of fundamental particles.**

1.8.2 What Determines Mass?

1. The **mass of any particle is simply the total mass of the Unit Particles within it.**
2. The **proton's mass is greater than the electron's mass** because it is composed of more Unit Particles.
3. This explains why **particle masses are fixed and not arbitrary**—they directly depend on **integer counts of Unit Particles.**

By eliminating arbitrary mass values and explaining charge through structural arrangements, UPT provides **a more fundamental understanding of matter.**

1.9 The Connection Between Unit Particles and Space-Time

If **Unit Particles are the only fundamental entities**, how do they relate to space-time itself? UPT proposes that:

1. **Space is not an empty void—it is a structured medium formed by a vast number of Unit Particles.**
2. **The density of Unit Particles determines the properties of space, including gravity and relativity effects.**
3. **Time is a function of how Unit Particles interact and change configurations.**

This suggests that **space and time are emergent properties of Unit Particles**, rather than fundamental entities. This perspective aligns with recent ideas in quantum gravity that suggest space-time arises from deeper structures.

1.10 Key Takeaways from Unit Particle Theory

1. **The Unit Particle is the only truly fundamental particle**—all others are structured groupings of it.
2. **Electrons, protons, and neutrons are formed from exact integer numbers of Unit Particles.**
3. Charge arises from how Unit Particles are

arranged, not from separate "charge carriers."
3. **Mass is simply the total number of Unit Particles in a particle.**
4. **Space and time emerge from the structured arrangement of Unit Particles.**

This **new perspective** simplifies the complex framework of modern physics, providing a **more unified and deterministic explanation of reality**.

Chapter 2: Unit Particle Theory vs. The Standard Model

2.1 Introduction: The Need for a New Theory

The **Standard Model of Particle Physics (SM)** has been the dominant theory explaining fundamental particles and forces. It describes how **quarks, leptons, and force carriers** interact through the electromagnetic, weak, and strong nuclear forces. However, despite its success, the Standard Model has **several critical shortcomings**:

1. **Too Many "Fundamental" Particles** – The Standard Model requires multiple types of quarks, leptons, and force carriers, with no clear explanation of why nature needs so many.
2. **No Explanation for Mass Origins** – While the Higgs mechanism explains mass generation, it does not reveal why particles have the specific masses they do.
3. **Exclusion of Gravity** – The Standard Model does not incorporate gravity, making it incompatible with **General Relativity**.

4. **Dark Matter and Dark Energy Remain Unexplained** – The universe appears to contain unknown mass and energy, yet the Standard Model provides no insights into their nature.
5. **Arbitrary Parameters** – The masses, charges, and force strengths of particles appear **manually inserted** into the theory rather than emerging from deeper principles.

Unit Particle Theory (UPT) offers **a radically different and simpler approach**:

1. **There is only one true fundamental particle—the Unit Particle (UP).**
2. **All known particles are structured compositions of Unit Particles in integer ratios.**
3. **Mass, charge, and force interactions emerge from the properties and configurations of Unit Particles.**
4. **UPT eliminates arbitrary constants by deriving particle properties from a single building block.**

In this chapter, we will **compare UPT with the Standard Model** and show why UPT provides a more **unified and logically consistent framework**.

2.2 Fundamental Particles: Many vs. One

2.2.1 The Standard Model's Complex Particle Zoo

The Standard Model describes **17 "fundamental" particles**:

1. **6 Quarks** (up, down, charm, strange, top, bottom)
2. **6 Leptons** (electron, muon, tau, and their neutrinos)
3. **4 Force-Carrying Bosons** (photon, W & Z bosons, gluon)
4. **1 Higgs Boson**

This complexity raises **several questions**:
1. Why does nature require **so many different fundamental particles**?
2. Why do quarks and leptons exist in **three generations** with no clear reason?
3. Are these particles truly fundamental, or are they composed of even smaller entities?

2.2.2 Unit Particle Theory's Simplicity

Unit Particle Theory proposes a **single, truly fundamental entity**:

1. **The Unit Particle (UP) is the only fundamental particle.**
2. **All observed particles (electrons, protons, neutrons, etc.) are structured formations of Unit Particles.**
3. **Each particle's mass and charge depend on the number and arrangement of Unit Particles within it.**

This eliminates the need for **multiple quarks, leptons, and force carriers**, replacing them with **a single universal building block**.

2.3 The Origin of Mass: Higgs Field vs. Unit Particle Composition

2.3.1 The Standard Model's Higgs Mechanism

In the Standard Model, mass arises from interactions with the **Higgs field**. The Higgs boson, discovered in 2012, validates this mechanism but does **not explain why particles have their specific masses**.

2.3.2 UPT's Explanation: Mass as a Count of Unit Particles

UPT provides a **more straightforward** mass-generation mechanism:

1. **Mass is simply the total mass of the Unit Particles in a particle**.
2. **The proton is heavier than the electron because it contains more Unit Particles**.
3. **This naturally explains why particle masses are fixed and why they follow precise integer ratios**.

Instead of relying on **an abstract Higgs field**, UPT directly connects mass to the number of fundamental building blocks.

2.4 The Nature of Charge: Arbitrary Property vs. Emergent Structure

2.4.1 Standard Model: Charge as an Intrinsic Property

In the Standard Model, charge is a **fundamental, unexplained property**. Particles like electrons and protons **simply "have" charge** with no deeper reason.

2.4.2 UPT: Charge as a Structural Phenomenon

UPT suggests that **charge arises from the spatial organization of Unit Particles**:

1. **Electrons** have a specific Unit Particle configuration that generates **negative charge (-1e)**.
2. **Protons** have a different structure that produces **positive charge (+1e)**.
3. **Neutrons** have a balanced arrangement, leading to **neutrality**.

Charge is not an arbitrary characteristic—it is a **direct result of how Unit Particles arrange themselves in space**.

2.5 The Nature of Forces: Standard Model's Four vs. UPT's Unified Interaction

2.5.1 The Standard Model's Four Separate Forces

The Standard Model requires four **fundamental forces** to explain interactions:

 1. **Electromagnetism** (mediated by photons)
 2. **Strong Nuclear Force** (mediated by gluons)

3. **Weak Nuclear Force** (mediated by W & Z bosons)
4. **Gravity** (not explained by the Standard Model)

These forces operate **independently** with no clear unification.

2.5.2 UPT: Forces as Interactions of Unit Particles

UPT proposes a **single underlying mechanism**:
1. Forces emerge from the way Unit Particles interact.
2. The "force" between particles depends on the density, arrangement, and movement of Unit Particles.
3. Gravity, electromagnetism, and nuclear forces are all manifestations of Unit Particle configurations.

This **unifies fundamental interactions** instead of treating them as independent forces.

2.6 Why Unit Particle Theory Is a Better Model

2.6.1 Solving the Problems of the Standard Model

UPT resolves key weaknesses of the Standard Model:

Problem with Standard Model	Solution in Unit Particle Theory
Too many fundamental particles	Only one: the Unit Particle

Problem with Standard Model	Solution in Unit Particle Theory
Mass requires the Higgs field	Mass = Total number of Unit Particles
Charge is arbitrary	Charge = Structural property of Unit Particles
Forces are separate entities	Forces emerge from Unit Particle interactions
No explanation for gravity	Gravity arises from Unit Particle density
Cannot explain dark matter	Dark matter = Unobservable states of Unit Particles
Cannot explain dark energy	Dark energy = Unit Particle density variations

2.6.2 Why UPT is a More Unified Theory

1. Simplifies physics by reducing everything to a single fundamental particle.
2. Eliminates arbitrary constants and unexplained properties.
3. Unifies mass, charge, and forces under one framework.
4. Provides a more deterministic and structured foundation for physics.

2.7 Conclusion: The Need for a Paradigm Shift

The Standard Model has been incredibly successful, but it is **not the final answer**. It remains incomplete and filled with unresolved questions. **Unit Particle Theory offers a more fundamental, unified, and elegant framework for understanding reality.**

Chapter 3: Unit Particle Theory and the Nature of Space-Time

3.1 Introduction: The Relationship Between Matter and Space-Time

Classical physics and modern theories like **General Relativity (GR)** and **Quantum Mechanics** treat **space and time as fundamental entities**. In Einstein's relativity, space-time is a continuous fabric that can bend under the influence of mass and energy. However, in **Quantum Field Theory (QFT), space-time is a background stage where quantum interactions occur**.

But what if **space-time is not fundamental**?

Unit Particle Theory (UPT) proposes that:

1. **Space is not an empty void—it is a structured medium formed by Unit Particles.**
2. **Time is not absolute but emerges from the dynamic interactions of Unit Particles.**
3. **Gravity, energy, and motion arise naturally from the density and arrangement of Unit Particles in space.**

This chapter explores **how Unit Particles create space-time, gravity, and energy interactions.**

3.2 Is Space Truly Empty? The Unit Particle View

3.2.1 The Standard View of Space
In classical physics, space is considered:
1. **A passive, empty stage** where objects move.
2. **Absolute (Newtonian physics) or curved by mass-energy (Einstein's relativity).**
3. **Independent of matter (except in relativity, where mass-energy affects it).**

3.2.2 UPT: Space as a Structured Medium
UPT suggests that **space is not truly empty**—it consists of a vast distribution of Unit Particles:
1. **Unit Particles exist everywhere, forming the fabric of space itself.**
2. **The density and arrangement of Unit Particles determine the properties of space.**
3. **Motion through space is actually motion through a structured medium of Unit Particles.**

This means that instead of treating space as **an empty void or a curved mathematical construct**, UPT sees space as **a physical substance made of fundamental units**.

3.3 Time as an Emergent Property of Unit Particle Interactions

3.3.1 The Problem with Time in Physics

Physics does not explain **why time flows forward** or why different observers experience time differently (relativity).

In **Quantum Mechanics**, time is just a parameter.
In **General Relativity**, time is distorted by gravity.
But neither theory explains:
1. **Why time exists at all.**
2. **Why time flows in one direction.**
3. **Why time slows down near massive objects.**

3.3.2 UPT: Time as a Function of Unit Particle Motion

UPT proposes that **time is not fundamental—it emerges from the movement and interactions of Unit Particles**:
1. **Time is a measure of changes in the configuration of Unit Particles.**
2. **Faster movement through a dense Unit Particle field results in time dilation.**
3. **Near massive objects (where Unit Particles are more densely packed), changes occur more slowly, explaining gravitational time dilation.**

Thus, **time is simply a measure of how Unit Particles interact and change their positions in space.**

3.4 Gravity as an Emergent Effect of Unit Particle Density

3.4.1 The Standard Model's View on Gravity

Gravity is currently explained by **Einstein's General Relativity**, where:

1. Mass bends space-time, and objects move along curved paths.
2. Gravitational waves propagate disturbances in space-time.
3. Quantum Gravity remains unsolved (no explanation for gravity at a quantum scale).

However, Einstein's model does **not** explain:
1. What space-time is made of.
2. Why mass bends space.
3. Why gravity is so much weaker than the other forces.

3.4.2 UPT: Gravity as a Result of Unit Particle Interactions

UPT suggests a radically different explanation:
1. **Gravity is not a separate force—it arises from the density of Unit Particles in space.**
2. **Massive objects contain a high concentration of Unit Particles, which causes surrounding Unit Particles to cluster more tightly.**
3. **This clustering effect creates an attraction, which we perceive as gravitational pull.**

In this view, **gravity is not a "force" but an emergent property of how Unit Particles distribute themselves around massive objects**.

This theory naturally explains:
1. **Why gravity is weaker than electromagnetism**—it is not a force but a density-based effect.
2. **Why time slows down near massive objects**—higher Unit Particle density reduces the rate of change in configurations.
3. **Why gravitational waves exist**—disturbances

in the Unit Particle field propagate through space.

Thus, **space, time, and gravity are all emergent effects of the fundamental Unit Particle medium.**

3.5 Motion and Energy in the Unit Particle Framework

3.5.1 What Is Energy in Standard Physics?

Energy is treated as an abstract property that comes in various forms:
1. **Kinetic energy (motion).**
2. **Potential energy (stored energy).**
3. **Heat energy, electrical energy, etc.**

However, physics **does not explain what energy fundamentally is**—it is merely a mathematical quantity.

3.5.2 UPT: Energy as Unit Particle Rearrangement

UPT suggests that **energy is a measure of how Unit Particles move and interact**:

1. **Motion = The rearrangement of Unit Particles in space.**
2. **Kinetic Energy = The relative motion of Unit Particles.**
3. **Potential Energy = Stored tension within the Unit Particle field.**
4. **Heat = The random vibrational movement of Unit Particles.**

This means that energy is **not a separate property**—it is a direct consequence of **Unit Particle movement.**

3.6 Dark Matter and Dark Energy: A Unit Particle Perspective

3.6.1 The Mystery of Dark Matter

Dark Matter is **invisible** and **interacts gravitationally but not electromagnetically**. The Standard Model cannot explain it.

UPT's explanation:
1. **Dark Matter = A unique configuration of Unit Particles that does not emit or absorb light.**
2. **It still influences gravity because it increases Unit Particle density in space.**

3.6.2 The Nature of Dark Energy

Dark Energy causes the universe's expansion to accelerate. The Standard Model offers no reason why.

UPT's explanation:
1. **Dark Energy = A decrease in the density of Unit Particles over large scales, reducing gravitational attraction.**
2. **This leads to an apparent acceleration in cosmic expansion.**

Thus, UPT naturally accounts for both **dark matter and dark energy without introducing new exotic particles.**

3.7 Summary: The New View of Space-Time

Conventional View	Unit Particle Theory (UPT)

Conventional View	Unit Particle Theory (UPT)
Space is a passive void	Space is a structured medium of Unit Particles
Time is fundamental	Time emerges from Unit Particle interactions
Gravity is a force	Gravity is an effect of Unit Particle density
Energy is an abstract concept	Energy is Unit Particle movement
Dark Matter is mysterious	Dark Matter is an unseen state of Unit Particles
Dark Energy is unexplained	Dark Energy is a decrease in Unit Particle density

3.8 Conclusion: A New Foundation for Physics

UPT offers a profound **shift in our understanding of space-time, gravity, and energy**:
1. **Space is not empty—it is made of Unit Particles.**
2. **Time is an emergent property of Unit Particle interactions.**
3. **Gravity results from the density of Unit Particles, not from space-time curvature.**
4. **Energy is simply the movement and rearrangement of Unit Particles.**

5. **Dark Matter and Dark Energy naturally arise from different states of Unit Particles.**

This **unifies physics into a single framework**, providing deeper insights into **the true nature of reality**.

3.9 The Physical Structure of Space in UPT

3.9.1 Is Space Truly Continuous?

Traditional physics treats space as either:
1. **A continuous geometric entity (General Relativity)**, or
2. **A discrete but unobservable "quantum foam" (Quantum Gravity speculations)**..UPT, however, suggests that:
3. **Space itself is composed of an invisible but structured medium of Unit Particles.**
4. **These Unit Particles act as the fundamental "fabric" of space.**
5. **Space does not exist independently—it is a result of the distribution of Unit Particles.**

This challenges the classical idea that **space is just an empty container** and instead proposes that **what we perceive as "space" is actually a structured arrangement of Unit Particles.**

3.10 How Gravity and Inertia Arise from Unit Particle Density

3.10.1 Why Do Objects Resist Motion?

Inertia—the resistance of objects to changes in motion—has always been treated as an intrinsic property of mass. But **why does mass resist acceleration?**

UPT proposes:
1. **Inertia results from an object interacting with the surrounding Unit Particle field.**
2. **The denser the Unit Particle environment, the greater the resistance to motion.**
3. **Inertia is not a separate property—it emerges naturally from how an object's internal Unit Particles interact with the external Unit Particle medium.**

This **solves a long-standing mystery**:
1. Why does every object, regardless of composition, experience the same inertia?
2. Why does inertia seem to increase in high-energy environments (relativistic mass increase)?

UPT suggests that inertia is **not an inherent mass property but an effect of moving through the structured field of Unit Particles.**

3.10.2 Why Does Gravity Exist?

Einstein's General Relativity explains gravity as **the curvature of space-time** but does **not explain what space-time is made of**.

UPT provides a more **mechanistic** explanation:
1. **Massive objects contain large numbers of Unit Particles, creating a high-density field around them.**
2. **Nearby Unit Particles naturally redistribute themselves, creating a density gradient.**
3. **This density variation leads to what we perceive as gravitational attraction.**

1. In essence, **gravity is not a fundamental force but a result of how Unit Particles distribute themselves around mass.**
   1. **Explains why gravity is weaker than the other forces**—it's an emergent property, not a force in itself.
   2. **Explains why gravity is always attractive**—it's a result of density variations, not charge-based interactions.

3.11 Why Does Time Slow Down in High-Density Regions?
3.11.1 The UPT Explanation of Time Dilation
In General Relativity:
Time slows down near massive objects due to space-time curvature.
Time also slows for fast-moving objects due to relativistic effects.
But **why does this happen?**
UPT provides a direct physical explanation:
1. **Time is a measure of change within the Unit Particle field.**
2. **When Unit Particle density increases (such as near a massive object), changes happen more slowly.**
3. **Thus, the "flow of time" is a function of Unit Particle density**—not space-time curvature.

This **clarifies the true nature of time dilation**:
1. **Near black holes, Unit Particle density is extremely high, so time slows significantly.**
2. **At high speeds, an object moves through the**

Unit Particle field at a different rate, changing its interaction with time.

3.12 The Nature of Motion in UPT

3.12.1 How Do Objects Move in a Unit Particle Medium?

In classical physics, motion is explained using **Newton's Laws**:
1. An object stays in motion unless acted upon by a force.
2. Force = Mass × Acceleration.

But **what is happening at a fundamental level?**
UPT explains motion as **a continuous rearrangement of Unit Particles**:
1. **An object moves by transferring momentum through the surrounding Unit Particle medium.**
2. **Higher densities of Unit Particles create resistance, which we perceive as inertia.**
3. **A force is simply an external interaction that alters the Unit Particle distribution within an object.**

This provides a **new interpretation of momentum and kinetic energy**:
1. **Momentum is not just "mass times velocity"**—it is a measure of how many Unit Particles are in motion relative to the field.
2. **Acceleration occurs when the internal structure of Unit Particles shifts to a new state of equilibrium.**

3.13 Why a Graviton Is Not Needed in UPT

Quantum Gravity theories try to **quantize gravity** by introducing a hypothetical particle—the **graviton**. However:
1. The graviton has never been detected.
2. **It does not naturally fit into existing quantum field theories.**

UPT **removes the need for a graviton**:
1. **Gravity is not a fundamental force—it is an emergent effect of Unit Particle density gradients.**
2. Since it is not a force, it does not require a force carrier particle (like photons for electromagnetism).
3. Gravitational waves are simply disturbances in the Unit Particle medium, not the result of graviton exchange.

This solves the biggest problem in modern physics—the conflict between Quantum Mechanics and General Relativity—without the need for hypothetical particles.

3.14 Summary of Additional Concepts

Traditional View	Unit Particle Theory (UPT)
Space is an empty void	Space is a medium of Unit Particles
Inertia is a property of mass	Inertia results from resistance in the Unit Particle field

Traditional View	Unit Particle Theory (UPT)
Gravity is a force	Gravity is an emergent effect of Unit Particle density
Time is fundamental	Time emerges from Unit Particle interactions
Motion is a fundamental property	Motion is a continuous rearrangement of Unit Particles
Gravitons are needed	Gravity does not require a particle carrier

3.15 The Future of Unit Particle Theory

The **implications of UPT** go far beyond current physics:
1. **It offers a framework to unify quantum mechanics and gravity.**
2. **It provides a structured explanation for dark matter and dark energy.**
3. **It challenges the Standard Model's reliance on multiple fundamental forces.**
4. **It may lead to new breakthroughs in energy, space travel, and quantum computing.**

By **replacing the abstract concept of space-time with a real, structured field of Unit Particles**, we open the door to a **new era of physics**—one that is simpler, more intuitive, and deeply connected to fundamental reality.

Chapter 4: Unit Particle Theory and Quantum Mechanics

4.1 Introduction: The Mystery of Quantum Mechanics

Quantum Mechanics (QM) is **one of the most successful theories in physics**, accurately describing atomic and subatomic phenomena. However, it also presents **deep paradoxes and unresolved mysteries**, such as:

1. **Wave-Particle Duality:** How can particles behave both as discrete objects and as waves?
2. **Quantum Entanglement:** How can two particles remain instantaneously connected across vast distances?
3. **Quantum Superposition:** How can a particle exist in multiple states until observed?
4. **Heisenberg Uncertainty Principle:** Why does measuring one property (like position) disturb another (like momentum)?

The **Standard Model** explains these phenomena mathematically but **does not provide a physical mechanism** for why they happen.

Unit Particle Theory (UPT) proposes that all quantum effects arise naturally from the behavior of the fundamental Unit Particles.

This chapter will explain how **quantum mechanics can be derived from UPT** and why it offers a more intuitive understanding of reality.

4.2 Wave-Particle Duality Explained by Unit Particles

4.2.1 The Classical View of Wave-Particle Duality

In the early 20th century, scientists discovered that **light, electrons, and other particles behave both as waves and particles**.

1. In the **double-slit experiment**, electrons sometimes behave like particles (creating two bands) and sometimes like waves (creating an interference pattern).
2. In **quantum field theory**, particles are described as "quantum fields" that spread over space like waves.

But **why does a single particle act like a wave?**

4.2.2 The UPT Explanation: Unit Particles as the Medium of Waves

UPT proposes that:

1. **All fundamental particles (electrons, protons, photons, etc.) are formed from clusters of Unit Particles.**
2. **Unit Particles create an invisible "quantum medium"** through which these particles propagate.

3. What we call a "wave" is the disturbance of the surrounding Unit Particle field as a fundamental particle moves.

This solves the wave-particle duality mystery:
1. The particle is real—it consists of Unit Particles.
2. The wave is also real—it is a disturbance in the surrounding Unit Particle medium.
3. This explains why a single electron "interferes with itself"—it is moving through a structured medium that affects its path.

Thus, **the wave-like behavior of quantum particles is not an inherent property of the particle itself but a result of how it interacts with the surrounding Unit Particles.**

4.3 Quantum Superposition and the Unit Particle Framework

4.3.1 The Mystery of Superposition

In QM, **a particle can exist in multiple states simultaneously until measured.** For example:
1. An **electron** in an atom does not have a fixed location until observed.
2. A **photon** can take multiple paths at once until detected.

How can a physical object exist in multiple places at once?

4.3.2 UPT Explanation: Superposition as a Probabilistic Effect of the Unit Particle Field

UPT suggests that:

1. The "multiple states" of a particle are actually different possible interactions with the surrounding Unit Particle medium.
2. Until an external interaction (measurement) forces a definite state, the particle's future position remains probabilistic.
3. Superposition is not the particle itself existing in multiple places, but the potential paths it could take based on how the Unit Particle field is disturbed.

This explains why **measurement collapses the wave function**—the act of measuring interacts with the Unit Particle field, forcing the system into a definite configuration.

4.4 Quantum Entanglement: A Unit Particle Connection

4.4.1 The Classical View of Entanglement

Quantum entanglement describes how two particles remain **instantaneously correlated** even when separated by vast distances.

1. If one entangled electron's spin is measured as "up," the other will be "down"—**no matter how far apart they are.**
2. This appears to violate Einstein's relativity, which states that no information can travel faster than light.

4.4.2 The UPT Explanation: A Persistent Connection Through the Unit Particle Medium

UPT proposes:

1. Entangled particles remain connected through the Unit Particle field.
2. Instead of "transmitting" information, the state of one particle instantaneously affects the distribution of Unit Particles surrounding both particles.
3. The effect is non-local because Unit Particles form an underlying fabric that spans the entire universe.

This removes the paradox of faster-than-light communication because **information is not "sent" between particles—it already exists within the continuous Unit Particle field.**

4.5 Heisenberg's Uncertainty Principle in UPT

4.5.1 The Classical Interpretation

The **Heisenberg Uncertainty Principle** states that:
1. The more precisely we measure a particle's **position**, the less precisely we can know its **momentum** (and vice versa).
2. This is usually explained as an inherent property of quantum systems.

4.5.2 The UPT Explanation: Measurement Disturbs the Unit Particle Field

UPT explains uncertainty as:
1. **Particles are surrounded by a dense Unit Particle field.**
2. **Measuring position requires interacting with this field, which disturbs the particle's momentum.**

3. Measuring momentum affects the surrounding field, making position less certain.

Thus, **uncertainty is not a fundamental limit of nature but a consequence of interacting with the Unit Particle medium.**

4.6 The Collapse of the Wave Function in UPT

Quantum Mechanics describes **wave function collapse** as:
1. A particle exists in multiple states (superposition).
2. When measured, it "chooses" one definite state.

UPT suggests:

1. **The wave function is actually a real physical structure in the Unit Particle field, not just a mathematical tool.**
2. Observation disturbs the Unit Particle medium, causing the particle to settle into one possible path.
3. Collapse is not random but dictated by how the Unit Particles interact at the moment of measurement.

This provides a physical explanation for wave function collapse, removing the mystery of "why measurement matters" in quantum physics.

4.7 Summary: How UPT Resolves Quantum Mysteries

Quantum Phenomenon	UPT Explanation

Quantum Phenomenon	UPT Explanation
Wave-Particle Duality	A particle is real, but its wave behavior is due to interactions with the Unit Particle field.
Superposition	Multiple possible paths exist in the Unit Particle medium until one is selected.
Entanglement	Particles remain linked through the Unit Particle field, not faster-than-light signals.
Uncertainty Principle	Measuring disturbs the Unit Particle medium, affecting other properties.
Wave Function Collapse	Collapse occurs when the Unit Particle field is forced into a definite configuration.

4.8 Conclusion: The New Quantum Reality

Unit Particle Theory provides **a physical basis for quantum mechanics**, solving long-standing mysteries:

1. **It explains why particles behave like waves—because they move through a structured medium.**
2. **It removes the paradox of entanglement—**

particles remain linked via the continuous Unit Particle field.
3. It clarifies wave function collapse—measurement is an interaction with the fundamental fabric of space.

This **unites quantum mechanics with a deeper physical reality**, paving the way for **a new understanding of the universe.**

4.9 The True Nature of a Wave Function in UPT

4.9.1 The Traditional View: A Mathematical Tool?

In conventional quantum mechanics, the **wave function (ψ)** describes the probability of a particle's location and properties. However:
1. The wave function is often **considered abstract**, with no physical reality.
2. Some physicists argue it's just **a calculation tool**, not an actual structure in space.

4.9.2 The UPT View: A Real, Physical Structure in the Unit Particle Field

UPT proposes that:
1. The **wave function is not abstract**—it represents the actual deformation of the Unit Particle field.
2. Just like a sound wave is a real disturbance in air, the **wave function is a real disturbance in the underlying Unit Particle medium.**
3. The reason we get **probabilities in QM** is because interactions between Unit Particles are discrete, leading to statistical outcomes.

This resolves **a major debate in quantum mechanics**:
1. **The wave function is not just an equation—it's a real physical entity shaped by the fundamental Unit Particle medium.**
2. **This explains why quantum effects emerge even when we don't observe them—because the Unit Particle field is still evolving naturally.**

4.10 How UPT Explains Schrödinger's Equation

4.10.1 The Problem: What Does Schrödinger's Equation Really Mean?

Schrödinger's equation describes how the wave function evolves over time. However, standard quantum mechanics **does not explain why this equation takes the form it does.**

4.10.2 The UPT Explanation: Motion of Waves in a Structured Medium

UPT proposes that:
1. Schrödinger's equation naturally emerges when you describe how **disturbances (particles) move through a dense Unit Particle medium.**
2. The mathematical form of the equation is similar to equations used for wave motion in fluids or elastic materials.
3. This suggests that **quantum waves are just real oscillations in the fundamental Unit Particle fabric.**

This means that:
1. **Schrödinger's equation is not arbitrary—it describes the physics of a real, structured**

medium.

2. The complex wave function represents actual movements of the fundamental Unit Particles, not abstract probability waves.

4.11 Why Does Measurement "Collapse" the Wave Function?

4.11.1 The Standard Quantum View: A Mysterious Observer Effect

In traditional quantum mechanics, when we measure a particle:

1. Its **wave function "collapses"**, choosing one specific state.
2. But QM **never explains why this happens—only that it does.**

4.11.2 The UPT Explanation: Measurement Disturbs the Unit Particle Field

UPT offers a natural explanation:

1. A quantum measurement means **physically interacting with the Unit Particle field**—forcing it into a definite arrangement.
2. Before measurement, the system's Unit Particles **exist in a superposition of possible arrangements**.
3. When measured, **the Unit Particle medium undergoes a forced reconfiguration**, selecting one state.

This removes the **mystery of "observer collapse"**:

1. **It's not consciousness or observation that collapses the wave function—it's the physical interaction with the Unit Particle field.**
2. **This explains why measurement always**

produces a definite result—because it restructures the local Unit Particle environment.

4.12 How Does UPT Explain Quantum Tunneling?

4.12.1 The Standard View: A Particle Magically Passes Through a Barrier

Quantum tunneling allows a particle to **pass through an energy barrier even if it doesn't have enough energy** to cross it normally.

1. This effect is essential in nuclear fusion and electronics.

2. But **classical physics cannot explain how a particle can "disappear" from one side and reappear on the other.**

4.12.2 The UPT Explanation: Unit Particles as a Continuous Medium

UPT suggests that:

1. A particle in quantum mechanics is **not a rigid object**—it is a **structured formation of Unit Particles interacting with the surrounding field.**

2. When approaching a barrier, the **Unit Particle wave structure does not fully "stop" but instead spreads slightly into the barrier.**

3. In some cases, **this small penetration allows the particle's wave structure to reform on the other side of the barrier.**

This explains quantum tunneling as a natural effect of how disturbances propagate in a structured Unit Particle medium.

1. There is no "teleportation"—only the natural movement of a wave-like structure through a deformable medium.

4.13 UPT's Implications for Quantum Field Theory (QFT)

4.13.1 The Standard Quantum Field Theory View

Quantum Field Theory (QFT) describes particles as **vibrations in abstract quantum fields** that fill space.

1. QFT successfully predicts particle interactions but **does not explain what the quantum fields actually are.**
2. The theory assumes that **each particle has its own unique quantum field** (electron field, quark field, etc.).

4.13.2 The UPT Explanation: One Universal Unit Particle Field

UPT proposes a **simpler framework**:
1. **Instead of separate fields for every type of particle, there is just one fundamental Unit Particle field.**
2. **Different particles (electrons, quarks, photons, etc.) are just different patterns or clusters of Unit Particles.**
3. **Quantum interactions arise naturally from how these patterns interact with the surrounding Unit Particle structure.**

This provides a **more intuitive explanation for quantum interactions**:

1. All fundamental particles are just structured formations within the same underlying Unit Particle field.
2. Forces (electromagnetic, strong, weak) arise from how these formations influence each other through Unit Particle interactions.
3. This eliminates the need for separate, independent quantum fields, unifying all of physics under one structure.

4.14 Summary of Additional Insights

Quantum Phenomenon	UPT Explanation
The Wave Function	A real physical structure in the Unit Particle medium, not just a probability function.
Schrödinger's Equation	Describes the movement of disturbances in the structured Unit Particle field.
Measurement Collapse	Occurs because measurement disturbs the local Unit Particle structure.
Quantum Tunneling	Happens when a particle's wave structure spreads through the Unit Particle medium.
Quantum Fields	Not separate entities—just different patterns in one universal Unit Particle

Quantum Phenomenon	UPT Explanation
	field.

4.15 Conclusion: The New Quantum Reality with UPT

1. UPT provides a physical basis for quantum mechanics that removes many of its mysteries.
2. It eliminates the need for separate quantum fields, replacing them with a single, unified Unit Particle field.
3. It explains wave function collapse, quantum tunneling, and measurement effects using real physical interactions.
4. It suggests that quantum mechanics is not "weird"—it is simply the natural behavior of a structured medium.
5. With this new foundation, we are now ready to explore how UPT explains the fundamental forces of nature!

Chapter 5: Unit Particle Theory and the Fundamental Forces of Nature

5.1 Introduction: Unifying the Forces of Nature

Physics currently recognises **four fundamental forces** that govern the universe:
1. **Gravitational Force** – the attraction between masses
2. **Electromagnetic Force** – governs electric and magnetic interactions
3. **Strong Nuclear Force** – binds protons and neutrons in the nucleus
4. **Weak Nuclear Force** – responsible for radioactive decay

Each force is treated as **a separate entity** with its own mediating particle (graviton, photon, gluon, W/Z bosons). These forces operate on **different principles**, which is why physics has struggled to **unify them** under a single theoretical framework.

UPT offers a different solution:
All forces emerge from the behaviour, density, and structure of Unit Particles.
This chapter explores how **Unit Particle Theory (UPT)** redefines the concept of force as a **natural consequence of fundamental particle arrangements**—removing the need for separate force-carrying particles and offering a more **coherent, unified model.**

5.2 Rethinking Force: From Mediators to Mechanics

In classical and quantum physics, a force is often described as either:
1. A **field** (like gravity or electromagnetism)
2. A **particle exchange** (like photons for EM, gluons for the strong force)

But UPT asks:
What if force is not a transmission or exchange—but a **response to local conditions** in the Unit Particle field?

In UPT, all forces are manifestations of:
1. **Unit Particle density gradients**
2. **Interactions between clusters of Unit Particles**
3. **Reconfigurations within the Unit Particle medium**

Thus, **force is no longer a separate "thing"**—it is **a pattern of structural and positional relationships** among Unit Particles.

5.3 Gravity: A Density Effect, Not a Fundamental Force

As explored in earlier chapters:
1. In General Relativity, gravity is **spacetime curvature caused by mass**
2. In UPT, gravity arises from **variations in Unit Particle density**

UPT's Gravity Model:

1. Mass = A large number of tightly packed Unit Particles
2. These clusters cause nearby Unit Particles to **arrange differently**, creating a **gravitational gradient**
3. This results in **movement of smaller particles toward the denser region**—experienced as gravity

Why this explanation is powerful:

1. It **eliminates the need for a graviton**
2. It **links gravity to matter's structure directly**, not abstract geometry
3. It naturally explains why **gravity is weakest**, since it is **not a fundamental interaction** but a **density-based emergence**

5.4 Electromagnetism: Unit Particle Alignment and Polarity

Electromagnetism is traditionally described as a **force mediated by photons**, operating between charged particles. But it offers **no deeper explanation for charge**.

In UPT:

1. **Charge is a property of how Unit Particles are arranged within a structure**
2. **Positive or negative polarity** emerges from the **geometry and spin orientation** of the Unit Particle configuration
3. Electromagnetic attraction or repulsion results from **field realignments in the Unit Particle medium** caused by these configurations

How this explains classic effects:
1. **Electric fields** = structural tensions in the surrounding Unit Particle fabric
2. **Magnetic fields** = rotational dynamics of Unit Particles within moving particles
3. **Light** = wave disturbances propagated through the Unit Particle medium

No need to assume massless photons as independent particles—they are **manifestations of energy waves in the Unit Particle field**

5.5 The Strong Nuclear Force: High-Density Structural Binding

In the Standard Model, the **strong force** binds quarks together via **gluons**, and also holds protons and neutrons inside atomic nuclei.

Yet:
1. Gluons have never been directly observed
2. The strong force is **immensely powerful at tiny ranges** but **drops off sharply**

UPT Interpretation:

1. The **strong force is the result of extremely dense, compact arrangements of Unit Particles**
2. When two clusters (like protons or neutrons) are close enough, **their Unit Particles overlap and interlock**, creating an **exceptionally strong cohesive bond**
3. Beyond a certain distance, this **interlocking effect vanishes**, explaining the rapid falloff of the force

Thus, **strong nuclear binding is a mechanical, geometric consequence of Unit Particle structure**—not a field or particle exchange.

5.6 The Weak Nuclear Force: A Structural Instability

The **weak force** causes processes like **beta decay**, where particles change type. In the Standard Model, it is mediated by **W and Z bosons**—very massive and short-ranged.

UPT Explanation:

1. The **weak interaction** is not a force, but a **restructuring of Unit Particles within a decaying particle**
2. A neutron, for example, is a **highly specific arrangement of Unit Particles**
3. Over time, this structure may become **unstable**, causing it to **reorganise into a proton**, emitting energy (and associated particles) as it does

No need for external mediators—the **transformation is internal** to the Unit Particle configuration. This makes **weak force phenomena self-contained** and logical within UPT.

5.7 All Forces as Patterns of Unit Particle Dynamics

Standard Force	Standard Explanation	UPT Explanation
Gravity	Curved spacetime, graviton (hypothetical)	Unit Particle density gradient around massive structures
Electromagnetism	Photons exchanged between charges	Structural polarity and motion of Unit Particles in a field
Strong Nuclear Force	Gluon exchange among quarks	Short-range cohesion of interlocking Unit Particle arrangements
Weak Nuclear Force	W/Z bosons, particle transformations	Instability and reconfiguration of internal Unit Particle patterns

5.8 Unification through Unit Particle Dynamics

One of the greatest goals in physics is to **unify all fundamental forces** into a single, coherent model—often referred to as a **"Theory of Everything"**.

The Standard Model requires:
1. Multiple types of particles
2. Multiple fundamental forces
3. Separate frameworks for gravity and quantum theory

UPT achieves unification by showing:

1. **All forces emerge from a single origin**—the **arrangement, density, and motion** of **one fundamental type of particle**
2. There are **no "fields" in space**, only **real physical interactions in a medium composed of Unit Particles**
3. This **naturally integrates gravity, quantum effects, and force interactions**

Unit Particle Theory is not a patch to existing physics—it's a new foundation that explains it all.

5.9 Conclusion: From Multiplicity to Unity

Modern physics explains the universe using **many types of particles and many types of forces**, each with their own complex properties and behaviours.

Unit Particle Theory simplifies this picture:

1. There is **only one fundamental particle**: the Unit Particle
2. Every known particle is a **structured grouping of Unit Particles**
3. All known forces are **emergent effects of Unit**

Particle interactions
4. This theory offers a **unified, mechanical, and intuitive understanding** of the universe

5.10 Force Carrier Particles: A UPT Perspective

In the Standard Model of particle physics, forces are mediated by specific particles known as **bosons**:
1. **Electromagnetic Force**: Mediated by **photons**.
2. **Strong Nuclear Force**: Mediated by **gluons**.
3. **Weak Nuclear Force**: Mediated by **W and Z bosons**.
4. **Gravity**: Hypothetically mediated by **gravitons**, though these remain theoretical.

UPT reinterprets these mediators not as independent entities but as manifestations of disturbances or reconfigurations within the Unit Particle field:
1. **Photons**: Represent oscillations or waves propagating through the Unit Particle medium, corresponding to electromagnetic interactions.
2. **Gluons**: Symbolize the cohesive interactions arising from high-density interlocking Unit Particle structures, accounting for the strong force.
3. **W and Z Bosons**: Reflect transient configurations during the reorganization of Unit Particles in processes like beta decay, indicative of the weak force.

4. **Gravitons**: In UPT, gravity emerges from density gradients in the Unit Particle field, negating the necessity for a separate graviton particle.

This interpretation aligns with the understanding that force carrier particles in the Standard Model are conceptual tools to describe interactions, which, in UPT, are direct consequences of Unit Particle dynamics.

5.11 Range and Strength of Forces: Insights from UPT

The four fundamental forces exhibit varying ranges and strengths:

1. **Gravity**: Weakest but with infinite range.
2. **Electromagnetism**: Stronger than gravity, also with infinite range.
3. **Strong Nuclear Force**: Strongest force but operates over a very short range, approximately the diameter of an atomic nucleus.
4. **Weak Nuclear Force**: Stronger than gravity but weaker than electromagnetism and the strong force; it operates at subatomic distances.

UPT explains these characteristics through the properties and interactions of Unit Particles:

1. **Gravity's Weakness and Infinite Range**: As gravity results from subtle density variations in the Unit Particle field, its effect is pervasive but weak, extending over vast distances.

2. **Electromagnetism's Strength and Range**: Arising from structural polarity and motion within the Unit Particle medium, electromagnetic interactions can propagate indefinitely through the field.
3. **Strong Force's Strength and Short Range**: The strong force is a consequence of tightly interlocking Unit Particle configurations that are effective only at extremely close proximities, explaining its short-range nature.
4. **Weak Force's Intermediate Characteristics**: The weak force emerges from internal reconfigurations of Unit Particles within subatomic particles, leading to interactions that are stronger than gravity but limited to short distances.

This framework provides a cohesive understanding of why each force behaves as observed, rooted in the fundamental properties of Unit Particles.

5.12 The Quest for Unification: UPT's Contribution

Physicists have long sought a **Unified Field Theory** to describe all fundamental forces within a single framework. While the Standard Model successfully unifies electromagnetism, the weak force, and the strong force, it struggles to incorporate gravity.

UPT offers a path toward unification by proposing that all forces are emergent properties of a single entity—the Unit Particle. By attributing the origin of all forces to the interactions and arrangements of Unit Particles, UPT eliminates the need for multiple fundamental fields and mediating particles. This approach aligns with the broader goal of unifying the fundamental interactions by reducing them to manifestations of a singular underlying reality.

5.13 Conclusion: A Unified Vision through UPT

Unit Particle Theory provides a comprehensive framework that reinterprets the fundamental forces as natural outcomes of Unit Particle dynamics. By offering explanations for force mediators, interaction ranges, strengths, and the potential for unification, UPT presents a paradigm that simplifies and unifies our understanding of the fundamental interactions governing the universe.

Chapter 6: Unit Particle Theory and the Cosmos

6.1 Introduction: A New Cosmological Paradigm

The mysteries of the universe—its origin, structure, and fate—are traditionally addressed by the fields of **cosmology** and **astrophysics**, rooted in models like the **Big Bang Theory** and **General Relativity**. These rely on abstract notions such as **spacetime expansion**, **dark matter**, and **dark energy**, many of which are inferred but not directly observed.

Unit Particle Theory (UPT), however, offers a **materialist and structural** perspective of the universe. It suggests that **everything—from galaxies to subatomic particles—is composed solely of Unit Particles**, whose arrangements and densities define the properties of matter, energy, and space itself.

6.2 Rethinking the Universe's Origin

Conventional View: The Big Bang

In the Standard Model of cosmology, the universe began as a **singular point** of infinite density and temperature that expanded rapidly. This model explains:

1. Cosmic Microwave Background Radiation (CMB)

2. Redshift of galaxies
3. Elemental abundances

But it raises unresolved questions:
1. What caused the Big Bang?
2. What existed before it?
3. Why is the universe fine-tuned for life?

UPT Perspective: Eternal Unit Particle Medium

Unit Particle Theory rejects the concept of a finite beginning. Instead:
1. The universe **has always existed** as a **boundless ocean of Unit Particles**
2. Regions of higher density formed structures over time through **natural clustering and alignment**
3. Matter, galaxies, and stars emerged from **self-organising dynamics** in the Unit Particle field

This is a **continuous evolutionary process**, not a singular explosive event. No need for an unnatural beginning, inflationary periods, or exotic singularities.

6.3 Cosmic Structures as Emergent Forms

Formation of Matter

UPT posits that matter forms when **positive whole-number groupings of Unit Particles** stabilise into specific structures, giving rise to particles like:
1. **Electrons** (x Unit Particles)
2. **Protons** (y Unit Particles, where $y > x$)

3. **Neutrons**, atoms, molecules, and beyond

This process scales naturally:
1. From **particles to atoms**
2. From **atoms to dust**
3. From **dust to stars and galaxies**

All are **configurations within the continuous Unit Particle medium**.

Gravitational Binding Without Spacetime Curvature

In contrast to General Relativity's curvature-based explanation of gravity, UPT suggests:
1. Mass causes **local increases in Unit Particle density**
2. These density gradients create **pressure differentials**, pulling surrounding matter inward
3. This explains **galactic rotation curves and clustering**, without invoking **dark matter**

6.4 Light, Redshift, and Cosmic Expansion

Light as Wave Propagation in the Unit Particle Medium

According to UPT:
1. Light is not a stream of photons but **a wave through the Unit Particle field**
2. Its speed is **determined by the rigidity and density** of the field
3. This explains why **the speed of light is constant** and why light behaves both as wave and particle

Redshift Without Expansion

In the standard view, Redshift is interpreted as evidence of galaxies receding due to the expansion of space. UPT provides an alternative:
1. As light travels great distances, **its wave loses energy** through subtle interactions with the Unit Particle medium
2. This energy loss appears as **Redshift**
3. No need for **expanding space or accelerating universe**

Thus, what we observe is **aging or diffusion of energy** in a material field—not Doppler shift from moving galaxies.

6.5 Reassessing Dark Matter and Dark Energy

Dark Matter in Standard Cosmology

Dark Matter is hypothesised to account for:
1. Missing mass in galaxies
2. Gravitational lensing
3. Cosmic structure formation

Yet, **it has never been directly detected**.

UPT Interpretation

What appears as gravitational influence of "dark matter" is simply:
1. Effects of **denser regions of Unit Particles** that are **not visible matter** but still exert gravitational pull

These regions **do not require exotic particles**—just variations in the universal medium's density.

Dark Energy and the Illusion of Acceleration

Cosmic acceleration is deduced from Type Ia supernova data. But in UPT:
1. **Energy disperses naturally** through the Unit Particle medium
2. The apparent acceleration is a **misinterpretation of Redshift and brightness attenuation**
3. No mysterious "dark energy" is needed

6.6 A Universe Without Boundaries

UPT envisions the cosmos not as **a ballooning spacetime** but as an **endless field** of Unit Particles, within which:
1. Matter evolves
2. Energy transfers
3. Structures form and decay cyclically

There is **no edge, no centre, and no origin point**—only **a constant dynamic interplay** of structure and dissolution.

6.7 Implications for the Fate of the Universe

Without cosmic expansion or heat death, UPT proposes a **sustainable universe**:
1. Structures form, decay, and reform cyclically
2. Unit Particles are **eternal, indestructible, and reusable**

3. The universe is **not progressing toward disorder**, but is a **living, breathing field of evolving order**

6.8 Conclusion: Cosmology Through the Lens of UPT

Unit Particle Theory revolutionises cosmology by offering:
1. A material foundation for space, time, and matter
2. A natural explanation for Redshift, structure formation, and gravity
3. An alternative to unverifiable concepts like dark matter and dark energy
4. A universe that is eternal, evolving, and structurally coherent

"The stars, galaxies, and voids are not in space; they are **shapes of space itself**, sculpted from the same indivisible particle." — *Manoranjan Ghoshal*

Certainly, let's delve deeper into the implications of **Unit Particle Theory (UPT)** on our understanding of cosmology, addressing key aspects such as the universe's origin, the nature of cosmic structures, and the phenomena of dark matter and dark energy.

6.1 Rethinking the Universe's Origin: A UPT Perspective

Challenging the Big Bang Paradigm

The **Big Bang Theory** posits that the universe originated from an infinitely dense singularity approximately 13.8 billion years ago, expanding and cooling to form the cosmos we observe today.

While this model explains several observations, it raises profound questions:
1. **Causality**: What triggered the Big Bang?
2. **Pre-Bang Conditions**: What existed before this event?
3. **Singularity Issues**: How do we reconcile the concept of a singularity with known physical laws?

These questions highlight the limitations of the Big Bang model in providing a comprehensive understanding of the universe's inception.

UPT's Eternal Framework

Unit Particle Theory offers an alternative by proposing an **eternal, unbounded universe** composed of Unit Particles. In this view:
1. **No Singular Beginning**: The universe didn't emerge from a singular event but has always existed as an infinite expanse of Unit Particles.
2. **Continuous Evolution**: Structures within the universe arise, evolve, and dissipate through the perpetual interactions and reconfigurations of these particles.

This perspective eliminates the need for a creation event, suggesting instead a cosmos that is in a state of constant flux and transformation.

6.2 Formation and Evolution of Cosmic Structures

Emergence of Matter

In UPT, matter is not a fundamental constituent but an emergent property of Unit Particle configurations:
1. **Particle Formation**: Stable arrangements of Unit Particles give rise to subatomic particles (e.g., electrons, protons).
2. **Atomic and Molecular Structures**: These particles further combine to form atoms and molecules, leading to the diverse matter observed in the universe.

This hierarchical assembly underscores the role of Unit Particles as the foundational building blocks of all physical entities.

Gravitational Dynamics Reinterpreted

Traditional physics attributes gravity to the curvature of spacetime caused by mass. UPT offers a different explanation:
1. **Density Gradients**: Variations in Unit Particle density create regions of higher and lower pressure.
2. **Attractive Forces**: These pressure differentials result in the gravitational attraction observed between masses.

This model provides a tangible mechanism for gravity, rooted in the physical properties of the Unit Particle medium.

6.3 Addressing Dark Matter and Dark Energy

Dark Matter: A UPT Interpretation

The elusive nature of **dark matter**, inferred from gravitational effects yet undetected directly, poses challenges to conventional physics. UPT suggests:
1. **Invisible Mass**: Regions of increased Unit Particle density that do not manifest as observable matter could account for the gravitational influences attributed to dark matter.
2. **No Exotic Particles Required**: This explanation negates the need for hypothesizing unknown particles, grounding the phenomenon in variations of a known medium.

This approach aligns with observations without invoking entities beyond the current empirical framework.

Dark Energy and Cosmic Expansion

The accelerating expansion of the universe is commonly attributed to **dark energy**. UPT offers an alternative viewpoint:
1. **Energy Dispersion**: Over vast distances, energy dissipates through interactions within the Unit Particle field, leading to observations interpreted as cosmic acceleration.
2. **Redshift Reconsidered**: The Redshift of distant galaxies may result from energy loss in light waves traversing the Unit Particle medium, rather than from space itself expanding.

This perspective challenges the necessity of dark energy, proposing instead a mechanism based on known physical interactions.

6.4 Implications for Cosmological Models

Adopting UPT necessitates a reevaluation of several cosmological concepts:

1. **Universe's Fate**: An eternal, self-regulating cosmos suggests a dynamic equilibrium, contrasting with scenarios of heat death or perpetual expansion.
2. **Cosmic Microwave Background (CMB)**: The CMB could be interpreted as residual thermal energy within the Unit Particle field, rather than as an echo of a primordial explosion.
3. **Structure Formation**: Galactic and larger-scale structures emerge naturally from the inherent tendencies of Unit Particles to form complex arrangements.

These reinterpretations offer a cohesive framework that addresses existing observations while providing new avenues for exploration.

6.5 Conclusion: A Unified Vision of the Cosmos

Unit Particle Theory presents a paradigm wherein the universe is an eternal, dynamic system of fundamental particles. This model provides coherent explanations for the origin and evolution of cosmic structures, the nature of gravitational forces, and the phenomena attributed to dark matter and dark energy. By grounding cosmology in the interactions of a single entity—the Unit Particle—UPT offers a unified and testable framework for understanding the universe.

Chapter 7: Implications of Unit Particle Theory (UPT) for Modern Physics and Technology

7.1 Rewriting the Foundations of Physics

From Abstract Fields to Material Substance

In conventional physics, forces and interactions are mediated by abstract fields—electromagnetic, gravitational, etc.—that operate within spacetime. UPT transforms this model by proposing that **all physical phenomena originate from the dynamics of a real, material substrate: the Unit Particle Field**.

Implications:

1. **Redefinition of Space**: Space is no longer empty vacuum but a dense, structured field of Unit Particles.
2. **Unification of Forces**: All forces—strong, weak, electromagnetic, gravitational—emerge from variations in Unit Particle density and arrangement.

3. **Time as Emergent**: Time is not a dimension but a measure of **change** in configurations within this universal particle field.

7.2 Quantum Mechanics Reconsidered

Quantum mechanics is rife with paradoxes: wave-particle duality, superposition, entanglement. UPT reinterprets these phenomena as **emergent behaviours of particle clusters in a continuous medium**.

Examples:
1. **Wave-Particle Duality**: The duality arises because particles (e.g., electrons) are **configurations within a dynamic field**. Their apparent wave nature is a result of oscillations in the surrounding Unit Particle medium.
2. **Quantum Tunnelling**: Explained as a reorganisation of the field allowing a particle to "appear" on the other side of a potential barrier due to fluctuations in Unit Particle density.
3. **Entanglement**: Arises from structural coherence between particles formed from common Unit Particle origins—allowing instant configuration updates across space due to the **non-local continuity** of the medium.

Implications:

1. Collapse of the wavefunction becomes **a local field rearrangement** rather than a metaphysical event.
2. Probabilities in quantum mechanics can be replaced by **material field tendencies** and structural dynamics.

7.3 Rethinking the Role of Constants

Fundamental constants like the speed of light (c), Planck's constant (h), and the gravitational constant (G) are **considered intrinsic properties of nature**.

UPT posits that these constants are **emergent from the physical properties of the Unit Particle field**, such as:

1. **Particle Density**
2. **Field Elasticity**
3. **Interaction Tensions**

This means that if the **local structure of the Unit Particle field** varies (due to mass distribution, energy, etc.), these constants might subtly change too—providing a new lens to investigate **variability in fundamental constants**.

7.4 New Technologies from UPT Principles

A new understanding of matter and forces opens the door to **revolutionary technologies**, including:

A. Inertial Field Manipulation

If gravity is a result of Unit Particle density gradients, then:
1. It may be possible to **generate local gravity wells or shields** by manipulating Unit Particle configurations artificially.
2. Space propulsion could become feasible by creating **asymmetric field distortions**, propelling crafts without reaction mass.

B. Energy Extraction and Conversion
UPT implies that:
1. **Unit Particle configurations hold energy** that can be released or absorbed during reformation processes.
2. This may allow **high-efficiency energy generation** or even access to new forms of energy via **field phase transitions**.

C. Matter Synthesis
1. Understanding the **exact Unit Particle counts and configurations** of fundamental particles could enable **custom synthesis of matter**, including exotic or programmable materials.

7.5 Redefining the Nature of Consciousness
Consciousness is typically studied in biology and neuroscience, but UPT opens a physicalist route:
1. The brain could be viewed as a **dynamic field processor**, where neurons interact not just electrically, but through **complex Unit Particle field harmonics**.

2. This supports the idea that **conscious awareness** arises not just from neural activity but from **structural coherence in the underlying particle field**.

This perspective may lead to:
1. **Field-based cognitive interfaces**
2. **New models of consciousness and artificial intelligence**
3. Enhanced understanding of **human intuition and perception**

7.6 Philosophical and Scientific Impact

Adoption of UPT would fundamentally alter several long-held beliefs:
1. **Materialism Revalidated**: Matter is once again primary; abstract entities like spacetime are derivative.
2. **Unified Science**: Physics, chemistry, and biology become interlinked via the dynamics of one basic entity—**the Unit Particle**.
3. **Elimination of Unverifiable Constructs**: Concepts like dark matter, dark energy, curved spacetime, and quantum indeterminacy may be replaced by **tangible structural principles**.

7.7 Conclusion: A Technological and Intellectual Revolution

Unit Particle Theory is not merely a theoretical framework—it is a **blueprint for the next scientific and technological revolution**.

1. It provides material explanations for formerly abstract phenomena.
2. It offers paths toward **unifying physics** and developing transformative technologies.
3. It reconnects science with a **rational, observable, and testable worldview**.

"In every wave, particle, and force lies the motion of the smallest unit of existence. To master the Unit Particle is to master the laws of nature itself." — *Manoranjan Ghoshal*

Chapter 8: Experimental Verification and Future Research Directions

8.1 The Need for Empirical Grounding

No theory in physics is complete without **experimental validation**. The strength of Unit Particle Theory (UPT) lies not only in its philosophical elegance but in its **capacity to be tested and falsified**.

This chapter outlines potential **experiments, observations, and research programmes** that could confirm or refute the propositions of UPT, offering a roadmap for its scientific adoption.

8.2 Experimental Predictions Unique to UPT

UPT diverges from both the Standard Model and General Relativity by proposing a **material substrate (the Unit Particle field)** underpinning all physical phenomena. This leads to several novel predictions:

1. Micro-Scale Gravitational Variability

UPT suggests that **gravitational attraction results from density gradients** in the Unit Particle field, not spacetime curvature. This implies:
1. In **laboratory-scale setups**, slight mass shifts or arrangements could cause **measurable variations** in gravitational pull, not predicted by Newtonian or Einsteinian gravity.
2. Precision gravimeters and torsion balance experiments could detect these **minute gravitational anomalies**.

2. Light Wave Energy Depletion Without Doppler Effect

If **Redshift** is due to energy loss in the Unit Particle medium rather than expansion of space:
1. Light passing through **high-density Unit Particle regions** should experience more Redshift than through lower-density regions.
2. Controlled **long-path light experiments in media** simulating Unit Particle density (e.g., Bose–Einstein condensates) may reveal Redshift **without any movement of the source**.

3. Non-local Interaction Signatures

Entanglement-like effects may be measurable between particles created from the **same Unit Particle configurations**, even **without traditional quantum coherence mechanisms**.
1. Experiments could test if **macroscopically separated systems retain correlations** due to their shared origin in field formations.

8.3 Suggested Experimental Setups

A. Field Resonance Detection
1. Develop detectors that can sense **minute fluctuations** or vibrations in the surrounding Unit Particle field.
2. These might detect **non-electromagnetic, non-thermal disturbances** during particle creation, decay, or high-energy interactions.

B. Particle Construction from Known Unit Counts
1. Attempt to assemble stable particle-like structures using **controlled high-density plasma or quantum condensates** that mimic Unit Particle configurations.

C. Variable Constants Testing
1. Use **ultra-precise atomic clocks** in varying gravitational and energy environments to detect any **minute shifts in the value of constants**, like Planck's constant or the fine structure constant.

8.4 Cosmological Observations for Support

CMB Interpretation Variance
1. Re-analyse cosmic microwave background data under UPT assumptions: instead of being a Big Bang remnant, the CMB could be **thermal background noise** of the Unit Particle field.

2. New sky surveys may uncover **fluctuation patterns** matching predicted Unit Particle wave interactions.

Large-Scale Structure Clustering
1. Simulations of cosmic structure formation using **Unit Particle density rules** may yield galaxy distributions that better match observed filamentary structures than dark matter models.

8.5 Theoretical and Computational Research
Modelling the Unit Particle Field
1. Build computer models simulating a **continuous field of discrete, indivisible Unit Particles**.
2. Study emergent behaviours: can particles, fields, forces, and motions be re-created from basic alignment and movement rules?

Field Dynamics and Wave Propagation
1. Develop equations governing **wave propagation, density shifts, and force fields** in the Unit Particle substrate.
2. These could serve as **new foundations** replacing both the Standard Model and General Relativity.

8.6 Collaborations and Interdisciplinary Approaches
Advancing UPT requires **cross-field cooperation**:
1. **Physicists** to test, model, and interpret.

2. **Engineers** to design ultra-sensitive instruments.
3. **Computer scientists** to simulate Unit Particle field dynamics.
4. **Philosophers of science** to explore implications and limitations of materialist frameworks.

8.7 A Call for Open Scientific Inquiry

UPT does not claim immediate replacement of existing theories. Instead, it **invites open investigation** and offers:

1. A materially grounded alternative to abstract and probabilistic interpretations.
2. A testable framework with clear, observable consequences.
3. A pathway toward a **unified theory of matter and interaction**.

"The true test of science is not in the comfort of consensus, but in the courage to explore anew." — *Manoranjan Ghoshal*

Chapter 9: Philosophical Reflections and the Future of Scientific Thought

9.1 The Historical Quest for the Fundamental

Human inquiry has always been driven by the question: **"What is the universe made of?"**

From ancient atomism to modern quantum field theory, this quest has evolved—but so have its complications. Each new framework has layered more abstraction upon reality: invisible fields, uncertain particles, curved spacetime.

Unit Particle Theory (UPT) reclaims a materialist simplicity: **that all things arise from a single, indivisible, eternal substance—the Unit Particle.**

9.2 Materialism Reaffirmed

Matter as Primary Reality

Whereas modern theories often begin with mathematical abstraction, UPT starts with **concrete substance**. It posits:

1. The universe is composed of **real, physical units**.

2. There are **no metaphysical infinities**—no pointlike singularities, no undefined vacuum fluctuations.
3. Every phenomenon is a result of **the interaction, arrangement, and movement of Unit Particles**.

This returns physics to a foundation where **empirical observation and logical coherence** define reality—not mysticism or speculation.

9.3 Reconnecting Physics with Intuition

Many today feel alienated from the language and logic of modern physics. Concepts like:
1. Particles being in two places at once,
2. Forces without mechanisms,
3. Space expanding faster than light,

…are intellectually fascinating, but **emotionally and rationally dissonant**.

UPT speaks in a language closer to intuition:
1. That **something real must exist to act**.
2. That **cause must precede effect**.
3. That **continuity, locality, and substance** are not illusions, but truths.

9.4 Time, Causality, and Determinism

UPT redefines time not as a fourth dimension but as **the record of change** in Unit Particle configurations.

Key implications:

1. **Causality is preserved**: All events follow physical interactions between definite entities.
2. **Determinism is revitalised**: While UPT allows for complexity and unpredictability, it rejects pure randomness as foundational.
3. **Time is emergent**: It arises from the evolving state of the field, not from any built-in cosmic clock.

This provides a **new metaphysical coherence**—a bridge between scientific explanation and philosophical clarity.

9.5 Ethics of Scientific Thought

With the adoption of UPT comes a broader ethical challenge: to **conduct science with clarity, honesty, and openness**. This includes:

1. Rejecting the seduction of unfalsifiable theories.
2. Prioritising explanations that are **understandable, observable, and teachable**.
3. Accepting that **truth may be simple**, even if it is not yet fully understood.

UPT is a call not just for new physics, but for **a new culture of inquiry**—one rooted in humility, curiosity, and reason.

9.6 A New Scientific Renaissance

The potential of UPT lies not only in reshaping physics but in **inspiring a unified scientific worldview.**

Future possibilities include:
1. A **materialist theory of mind** that links consciousness to physical coherence in Unit Particle fields.
2. A **reunification of sciences**, from physics to biology to sociology, under shared fundamental laws.
3. A new era of **technological progress** informed by deep understanding of matter at its most basic level.

UPT invites thinkers, experimenters, and dreamers alike to participate in a **21st-century renaissance of rational thought.**

9.7 Conclusion: A Theory for the Ages

Unit Particle Theory offers not just answers, but a new way to ask questions.
1. It simplifies without oversimplifying.
2. It grounds abstraction in substance.
3. It speaks not only to scientists, but to every curious mind that has ever looked up at the stars and wondered, *"What are we made of?"*

"Beyond the atom, beyond the wave, lies the unity of being—the Unit Particle. To understand it is to see the world anew."

— *Manoranjan Ghoshal*

Appendices

Appendix A: Mathematical Formulations of UPT

To provide a rigorous foundation for UPT, this appendix introduces the mathematical structures and equations that describe the behaviour and interactions of Unit Particles.

A.1 Unit Particle Field Equations

The dynamics of the Unit Particle field can be encapsulated in a set of field equations analogous to those in classical field theories. Let $\rho(r,t)$ represent the density of Unit Particles at position r and time t. The continuity equation governing the conservation of Unit Particles is:

$$\frac{\partial \rho}{\partial t} + \nabla . \rho v = 0$$

Where v(r,t) denotes the velocity field of Unit Particles.

A.2 Emergent Force Relations

Assuming that variations in Unit Particle density give rise to observable forces, the force F experienced due to a density gradient can be expressed as:

$$F = -k\nabla\rho$$

where k is a proportionality constant that relates the density gradient to the resultant force.

A.3 Wave Propagation in the Unit Particle Medium

Considering the Unit Particle field as a medium that supports wave propagation, the wave equation can be formulated as:

$$\frac{\partial^2 \varphi}{\partial t^2} = c^2 \nabla^2 \varphi$$

Where ψ(r,t) represents the wave function describing disturbances in the Unit Particle field, and c is the characteristic wave speed in this medium.

Appendix B: Glossary of Terms

1. **Unit Particle**: The fundamental, indivisible entity proposed as the building block of all matter and fields.
2. **Unit Particle Field**: The continuous, all-pervading field composed of Unit Particles, whose dynamics give rise to physical phenomena.
3. **Density Gradient**: A spatial variation in the density of Unit Particles, hypothesised to be the origin of fundamental forces.
4. **Emergent Properties**: Characteristics such as mass, charge, and spin that arise from specific configurations and movements of Unit Particles.
5. **Field Resonance**: A condition where disturbances in the Unit Particle field reinforce each other, potentially leading to the formation of stable particle-like structures.

Bibliography

1. **Ghoshal, S. N.** (1950). "An Experimental Verification of the Theory of Compound Nucleus." *Physical Review*, 80(6), 939–940. Link
2. **Heyde, K.** (1999). *Basic Ideas and Concepts in Nuclear Physics: An Introductory Approach* (3rd ed.). IOP Publishing.
3. **Krane, K. S.** (2008). *Introductory Nuclear Physics*. Wiley-India.
4. **Lilley, J.** (2006). *Nuclear Physics: Principles and Applications*. Wiley.
5. **Griffiths, D. J.** (2008). *Introduction to Elementary Particles*. Wiley.
6. **Thomson, M. A.** (2009). *Particle Physics*. University of Cambridge. Lecture Notes
7. **Costanzo, D.** (2024). "Units and Special Relativity." *PHY304 Particle Physics Lecture Notes*. Link
8. **LibreTexts**. (2024). "Particle Physics and Cosmology (Summary)." *University Physics III - Optics and Modern Physics*. Link

Final Compilation

The exploration of **Unit Particle Theory (UPT)** presents a paradigm shift in our understanding of the universe. By positing a single, fundamental entity—the Unit Particle—as the cornerstone of all physical reality, UPT offers a unified framework that bridges the gaps between existing theories and addresses longstanding paradoxes.

Through the detailed chapters, we have examined the theoretical foundations, mathematical formulations, experimental implications, and philosophical ramifications of UPT. The appendices have provided additional depth, offering mathematical rigor and clarifying terminology.

As with any scientific theory, the true test of UPT lies in empirical validation. The proposed experimental setups and observations aim to substantiate the claims of UPT, paving the way for potential technological advancements and a deeper comprehension of the cosmos.

In embracing Unit Particle Theory, we embark on a journey toward a more cohesive and intuitive understanding of nature, grounded in the simplicity of a singular foundational element.

Printed in Great Britain
by Amazon

ebb17070-1207-4cd9-b828-937c8fc2cdffR01